梁燕　編著

新手入廚系列

粒粒飯香

前言

米 飯是中國的傳統主食，一碗米飯，與五味調配，可以供給大部分所需營養。而不同種類的米有不同的功效。白米是滋補之物，含有人體必需的澱粉質、蛋白質、維他命及鈣、鐵等營養成分，可以提供人體所需的營養和熱量。糙米含能促進血液循環和腸道蠕動，有助預防便秘。而糯米則可溫補脾胃，對脾胃氣虛、常常腹瀉的人有治療效果。

沒時間煮飯卻又想吃得健康？本書為你搜羅受歡迎的米飯食譜，例如中式的豉汁排骨鳳爪飯、生炒糯米飯、福建炒飯、焗豬扒飯以及韓式的石頭鍋飯、台式的滷肉飯、泰式的菠蘿炒飯和葡式的葡國焗雞飯等，只要簡單的步驟便可，讓你在短時間內輕鬆做出豐富且美味的菜式。

目錄

看圖買材料做菜

Buy ingredients according to the pictures

白米
白米要完整，不要碎，變舊的米呈白點不要購買。
White rice
Complete, not broken;
do not buy old rice with
white spot.

珍珠米
大粒，圓潤，顏色白。
Pearl rice
Big, round, white in
colour.

糙米
完整，帶黃色或淺啡色。
Unpolished / Brown rice
Complete, slightly yellow
or light brown.

紅米
顏色比較紅，不要啞色。
Red rice:
Pretty red, not dim.

生菜
顏色翠綠，變黑的不要購買。

Lettuce
Green, should not buy if in dim / dark colour.

菠蘿
要有香味，但个要有酒味。

Pineapple
Smell good without wine aroma.

冬瓜
肉質要結實，不要鬆。

Winter gourd
Firm texture, not loosen / slacken.

番茄
紅色圓潤，表示多肉多汁。

Tomato
Red, round and juicy.

番薯
胭脂紅色的番薯比較香。

Sweet potato
Pretty red sweet potato gives better smell / aroma.

洋葱
完整，沒有花痕或凹痕。

Onion
Complete, no poor concave mark / stain.

蝦米
乾身，沒有濕氣。

Dried shrimp
Dry without moisture.

瑤柱
顏色金黃。

Dried scallop
Colour is golden brown.

臘肉
色澤要鮮，不要啞色。

Preserved pork
Colour is sharp / bright,
not dim.

冬菇
有香味，而乾身厚肉。

Dried black mushroom
Dry, thick, with good smell /
aroma.

火鴨
要有油光，不乾身。

Roasted duck
Greasy, not dry.

牛仔骨
顏色要紅，不要瘀黑。

Beef short rib
Colour is pretty red, not dim.

蜆
蜆殼開了的蜆表示已
死，不可買。

Clam
Do not buy any clam
with open shell.

雞蛋
啡色的雞蛋比白色的味道更佳。

Egg
Brown egg has better taste than
white egg.

買回來的材料怎樣處理？ What to do with the ingredients?

米 Rice

米放在米缸內，加數粒蒜頭可延長儲存期。

Put in rice container with several cloves of garlic to keep it longer.

臘味 Preserved pork

煮臘味前先用熱水洗去表面的塵和油分。

Rinse with hot water to remove dust and oil before cooking.

蜆 Clam

用水養 1 小時，令蜆吐出沙泥。

Soak in water for an hour to remove sand and mud.

菌類 Fungus

要汆水才可去除霉味。

Blanch to get rid of poor / musty smell.

① ② ③

汆水、飛水或拖水 Blanch	煮沸一鑊水，將食材放水中煮一會，然後用水沖淨。 Boil ingredients in hot water, rinse with water.
免治 Mince	將肉剁碎。 Chop to make minced meat.
白鑊 Stir-fry without adding oil	放鑊中，不下油炒。 Do not use oil for stir-frying.
切度 Section	切段。 Cut into sections.
幼切粗剁 Shred and chop	將肉切成薄片再切條，再切粒才剁，就可不用剁太久。 Slice meat and cut into strips, dice and chop to minimize chopping time.
粗粒 Coarse size	切為比較大粒。 Dice into large pieces.
幼粒 Fine size	在粗粒和剁碎之間。 Chop into pieces in between coarse and minced sizes.
茸 Purée	剁至最幼細。 Chop until become purée.
啤水 Rinse	在水喉下猛力沖水。 Rinse under plunging water.

煮飯的技巧 Cooking Techniques

蒸
Steam

蒸飯的米要先用水浸，撈起米時不用全部去清水分，只要連同部分水撈起，放蒸籠蒸便可。

Soak rice for a while before cooking. Do not drain away all water but keep some when cooking.

炒
Stir-frying

炒飯時不可將飯壓實，要炒鬆。

Do not press rice tightly during stir-frying. Keep rice loosen / slacken.

煲仔飯
Hot pot rice

煲好的飯不要即時揭蓋，要焗片刻才有香味。

Do not remove cover immediately after cooking hot pot rice. Cover for a while to give aroma.

洋葱

切洋葱前先將洋葱浸在水中，才不會刺激眼睛。

Onion

Soak onion in water before cutting to avoid eye irritation.

炒飯

要用隔夜的飯，因剛煮熟的飯黏性比較強，炒時會黏在一起。

Stir-fried rice

Use overnight rice for stir-frying. As using cooked rice immediately is more viscous and easily sticking together during stir-frying.

湯飯

不要將湯加入飯來煮，只要將飯放在碗中，然後將湯倒入飯內即可。

Soup rice

Do not cook rice with soup directly, just put rice in bowl and pour in soup.

糙米

顆粒較硬，要比平時煮飯的水多加 1/3，飯的軟硬度才會適中。

Brown rice

Brown grain is pretty hard. Add 1/3 additional water to cook to make rice softer.

做菜和味道的常用語

Common phrases of cooking and tastes

食物味道／形容	Taste / Adjective
熟	cooked
沒熟	raw
生／沒熟	uncooked
太鹹	too salty
太長時間	too long
不夠鹹	not salty enough
不夠甜	not sweet enough
香	aromatic
甜	sweet
酸	sour
苦	bitter
辣	spicy
鹹	salty

準備／煮菜方式	Preparation / Cooking Method
切片	slice
切長一點	cut longer
切短少少	cut shorter
切塊	cut into wedges
切粒	cut into dice
蒸	steam
炸	deep-fry
煎	shallow-fry
炒菜	stir-fried vegetables
焯菜	blanch vegetables
煲湯	cook soup
燜豬肉	stew pork

常用技巧

Common skills

切角 Cut into wedges	把材料移動，切成三角形。 Roll ingredient and cut into triangles.
骨牌 Domino	材料先切成長形，再修切成長方形。 Cut ingredient into long pieces, and then rectangles.
去衣 Peel off thin layer	把栗子放熱水中煮 1~2 分鐘，去掉栗子外皮。 Cook chestnuts into boiling water for 1-2 minutes and peel off thin layers.
去皮 Peel	用刨子削去材料外皮。 Peel skin of ingredient.
料頭 Side ingredients	泛指薑、葱、蒜、乾葱和辣椒，協助提升物料的香味。 Usually refer to ginger, spring onion, garlic, shallot and red chili which make ingredients more aromatic.

泡油 Blanch with oil	將材料放入八成滾油中 2~3 分鐘，取出瀝油。 Cook ingredient in 80% boiled oil for 2-3 minutes, take out and drain.
汆水（飛水） Blanch	將材料放入滾水中焯 2~3 分鐘，取出過冷。 Cook ingredient in boiling water for 2-3 minutes, take out and rinse with cold water.
白鑊 Wok without adding oil	沒有添加任何物料、醬料或油等，只是把鑊燒熱後直接下物料烘乾水分。 Without adding any ingredients, sauce or oil, dry ingredient in a heated wok directly.
爆香 Sauté	用少量油加熱，放入料頭略炒至出味的程序。 Slightly stir-fry seasonings till aromatic with some heated oil.

上海菜飯配豬扒

Vegetable Rice in Shanghai Style with Pork Chop

⊚ 材料 | Ingredients

豬扒 3 件
白菜 100 克
白飯 2 碗
薑米 1 茶匙

3 pcs pork chop
100g Chinese cabbage
2 bowls rice
1 tsp minced ginger

醃料 | Marinade

鹽 1/2 茶匙	1/2 tsp salt
胡椒粉 1/2 茶匙	1/2 tsp pepper
雞粉 1/2 茶匙	1/2 tsp chicken powder

做法 | Method

1. 白菜洗淨，瀝乾水分，剁碎。
2. 豬扒洗淨，用刀背剁鬆，加醃料拌勻。
3. 燒熱油鑊，下豬扒煎至兩面金黃熟透，盛起，切件。
4. 再燒熱油鑊，爆香薑米，加入白菜和少許鹽拌炒至軟，盛起。
5. 白飯放大碗中，加入白菜拌勻成菜飯，放上豬扒即成。

1. Rinse Chinese cabbage, drain and chop.
2. Rinse pork chop, hammer with the back of knife blade, then marinate.
3. Heat wok with oil, shallow-fry pork chop until both sides are golden brown, drain and chop into pieces.
4. Heat wok with oil again, sauté minced ginger, add Chinese cabbage and season with salt, stir-fry until tender and dish up.
5. Put rice in a large bowl, stir in Chinese cabbage to make vegetable rice. Top with pork chop and serve.

入廚貼士 | Cooking Tips

- 白菜可不用加入鑊中炒，直接加入米中一同煮。
- 菜飯可加入蝦米以增加味道和口感。
- You may cook Chinese cabbage with rice together instead of stir-frying.
- Dried shrimps can be added into rice to enhance flavour and texture.

地瓜飯

Sweet Potato Rice

4 人
Serves 4

25 分鐘
25 minutes

材料 | Ingredients

地瓜（番薯）600 克	600g sweet potato
糙米 300 克	300g brown rice
豬瘦肉 150 克	150g lean pork
葱 3 條	3 stalks spring onion
蝦米 2 湯匙	2 tbsps dried shrimps

做法 | Method

1. 糙米淘洗淨，用水浸 3 小時。
2. 其他材料洗淨。豬瘦肉切粒，蝦米浸泡。葱切去根部和尾部，切粒。地瓜去皮，切粒。
3. 燒熱油鑊，下蝦米爆香，加入豬瘦肉炒勻，再加入地瓜炒勻。
4. 糙米放電飯煲中，加適量水煮，飯將乾水時，加入爆過的材料，焗至所有材料熟透，灑上葱粒即成。

1. Rinse brown rice and soak for 3 hours.
2. Rinse other ingredients. Dice lean pork, soak dried shrimps. Remove root and end part from spring onion and dice. Peel sweet potato and dice.
3. Heat wok with oil, sauté dried shrimps, add lean pork and stir-fry. Then stir in sweet potato and stir-fry.
4. Cook brown rice in rice cooker with appropriate amount of water. When little water left, add in sauté ingredients and cook until well done. Sprinkle diced spring onion on top and serve.

入廚貼士 | Cooking Tips

- 可改用白米代替糙米，亦可用一半糙米一半白米。
- 地瓜可改為南瓜。
- You may use white rice to substitute brown rice, or use 1/2 portion of brown rice and 1/2 portion of white rice.
- Also, you may use pumpkin to substitute sweet potato.

雜菌十穀炊飯

Mixed Mushrooms with Rice with Ten Mixed Grains

材料 | Ingredients

十穀米 300 克
豬瘦肉 120 克
草菇 120 克
秀珍菇 120 克
冬菇 4 朵
葡萄乾 1 湯匙
蒜茸 1 茶匙

300g rice with 10 mixed grains
120g lean pork
120g straw mushrooms
120g oyster mushrooms
4 pcs dried black mushrooms
1 tbsp raisins
1 tsp minced garlic

醃料 | Marinade

生粉 1/2 茶匙
鹽 1/4 茶匙

1/2 tsp cornstarch
1/4 tsp salt

調味料 | Seasonings

麻油 1/2 茶匙
魚露 1/2 茶匙

1/2 tsp sesame oil
1/2 tsp fish sauce

做法 | Method

1. 十穀米用水浸 3 小時，淘洗淨。放電飯煲中加水煮熟，水要比平時煲飯的水多 1/3。
2. 豬瘦肉洗淨，切粒，加醃料拌勻。
3. 冬菇浸軟，去蒂，切絲。
4. 草菇、秀珍菇洗淨，汆水。
5. 燒熱油鑊，下蒜茸爆香，加入瘦肉粒炒熟，再加入冬菇、草菇、秀珍菇和葡萄乾拌勻。
6. 將所有材料放入飯內拌勻，加入調味料拌勻即可。

1. Soak rice with 10 mixed grains for 3 hours and rinse. Cook in rice cooker with appropriate amount of water. The amount of water should be 1/3 more than that of cooking white rice.
2. Rinse lean pork, dice and stir in marinade.
3. Soak dried black mushrooms, remove stems and shred.
4. Rinse straw mushrooms and oyster mushrooms, and blanch.
5. Heat wok with oil, sauté minced garlic, add lean pork and stir-fry until done. Then add dried black mushrooms, straw mushrooms, oyster mushrooms and raisins and stir well.
6. Mix rice with all ingredients, then add seasoning and stir well. Serve.

入廚貼士 | Cooking Tips

- 十穀米也可用來煲粥。
- You may use rice with 10 mixed grains to make congee.

雜菌飯番茄盅

Rice with Assorted Mushrooms in Tomato Pot

材料 | Ingredients

番茄 1~2 個
雜色飯 1/2 碗
鮮雜菌 50 克
松子仁 1~2 湯匙
煙肉 2 片
薄荷 1~2 片（裝飾）

1~2 tomatoes
1/2 bowl rice in different colours
50g assorted fresh mushrooms
1~2 tbsps pine nuts
2 slices bacon
1~2pcs mint leaves (for dressing)

2 人
Serves 2

25 分鐘
25 minutes

◯◯◯ 調味料 | Seasonings

日式燒汁 1 湯匙
鹽 1/2 茶匙
糖 1/2 茶匙

1 tbsp Japanese sauce
1/2 tsp salt
1/2 tsp sugar

◯◯◯ 做法 | Method

1. 材料洗淨。煙肉、雜菌分別切碎。松子仁以白鑊烘香。
2. 燒熱鑊，炒香煙肉，加入雜菌粒、雜色飯和松子仁炒勻，加入調味拌勻，盛起。
3. 番茄挖空，放入炒飯。
4. 焗爐以 190℃預熱 2~3 分鐘，放入番茄盅焗 10 分鐘，即成。

1. Rinse ingredients. Dice bacon and assorted mushrooms respectively. Toast pine nuts in a wok without adding oil.
2. Heat wok and stir-fry bacon until fragrant, add assorted mushrooms, rice in different colours and pine nuts and stir-fry well. Add seasonings and mix well, dish up.
3. Scoop out the flesh of tomatoes and put fried rice into it.
4. Preheat oven to 190 ℃ for 2~3 minutes, put tomatoes into the oven and bake for 10 minutes, serve.

入廚貼士 | Cooking Tips

- 不用番茄可改用三色椒，效果相若但味道不同。
- Tomatoes can be replaced by green, yellow and red bell peppers, effect is similar while taste is different.

菠蘿炒飯

Fried Rice with Pineapple

⬤⬤⬤ 材料 | Ingredients

蝦 12 隻
菠蘿 2 片
豬瘦肉 100 克
洋葱 1/2 個
雞蛋 1 隻
急凍紅蘿蔔粒、青豆、粟米粒各 20 克
白飯 3 碗

12 pcs shrimps
2 slices pineapple
100g lean pork
1/2 onion
1 egg
20g frozen carrot, green beans and
corn grains respectively
3 bowls of cooked white rice

醃料 | Marinade

胡椒粉 1/2 茶匙
生粉 1/2 茶匙
鹽 1/8 茶匙

1/2 tsp pepper
1/2 tsp cornstarch
1/8 tsp salt

調味料 | Seasonings

生抽 1 茶匙
老抽 1/2 茶匙
鹽 1/2 茶匙
糖 1/2 茶匙

1 tsp light soy sauce
1/2 tsp dark soy sauce
1/2 tsp salt
1/2 tsp sugar

做法 | Method

1. 材料洗淨。蝦去殼去腸，加一半醃料拌勻。
2. 豬瘦肉切粒，加另一半醃料拌勻。
3. 洋葱去衣，切塊。菠蘿切塊。
4. 紅蘿蔔粒、青豆、粟米粒汆水，盛起，再瀝乾水分。
5. 燒熱油鑊，下白飯和雞蛋炒鬆，盛起備用。
6. 再燒熱油鑊，爆香洋葱，下瘦肉粒和蝦拌炒，再加入紅蘿蔔粒、青豆、粟米粒炒勻，將白飯回鑊，加入調味料和菠蘿拌勻即可。

1. Rinse ingredients. Remove shells and intestines from shrimps and stir in 1/2 portion of marinade.
2. Dice lean pork and stir in 1/2 portion of marinade.
3. Peel onion and cut into pieces. Cut pineapple into pieces.
4. Banch carrot, green beans and corn grains and drain.
5. Heat wok with oil, add cooked rice and egg white, stir-fry until loosen. Set aside.
6. Heat wok with oil again. Sauté onion until fragrant, add lean pork and shrimps, then add carrot, green beans and corn grains. Return white rice to wok, add seasonings and pineapple. Stir-fry until well done. Serve.

入廚貼士 | Cooking Tips

- 菠蘿要在炒好其他材料時才下，味道才會甜。
- Add pineapple at last when all other ingredients are well done to save the sweet flavour.

雜錦冬瓜粒泡飯
Rice with Winter Gourd and Mixed Ingredients in Soup

材料 | Ingredients

冬瓜 300 克
豬瘦肉 100 克
冬菇 3 朵
白飯 3 碗
上湯 2 杯

300g winter gourd
100g lean pork
3 dried black mushrooms
3 bowls cooked white rice
2 cups chicken broth

3 人
Serves 3

20 分鐘
20 minutes

醃料 | Marinade

生粉 1 茶匙
鹽 1/4 茶匙
糖 1/4 茶匙
胡椒粉 1/4 茶匙

1 tsp cornstarch
1/4 tsp salt
1/4 tsp sugar
1/4 tsp pepper

調味料 | Seasonings

鹽 1/2 茶匙
雞粉 1/2 茶匙
胡椒粉 1/4 茶匙

1/2 tsp salt
1/2 tsp chicken powder
1/4 tsp pepper

做法 | Method

1. 材料洗淨。冬瓜去皮，去籽，切粒。
2. 豬瘦肉切粒，加入醃料拌勻。
3. 冬菇浸軟，去蒂，切粒。
4. 白飯保暖，放大碗內。
5. 燒熱油鑊，下冬菇粒略炒，加入瘦肉粒炒至熟，注入上湯待滾片刻，加入調味料，倒入飯內即成泡飯。

1. Rinse ingredients. Peel winter gourd, remove seeds and dice.
2. Dice lean pork and marinate.
3. Soak dried black mushrooms, remove stems and dice.
4. Keep cooked white rice warm and put into a large bowl.
5. Heat wok with oil. Stir-fry dried black mushrooms, add lean pork and cook until well done. Add broth and cook for a while, stir in seasonings, pour into rice to make rice in soup.

入廚貼士 | Cooking Tips
- 冬瓜不要切得太細，否則煮腍後會變成茸。
- Do not cut winter gourd into tiny pieces. Otherwise, it will become purée after cooking.

6 人
Serves 6

15 分鐘
15 minutes

Crispy Rice with
Assorted Mushrooms

雜菌鍋巴

材料 | Ingredients

鍋巴 1 包
本菇 1 包
金菇 1 包
西蘭花 1 棵
蒜茸 2 茶匙
薑茸 1 茶匙

1 packet crispy rice
1 packet hon shimeji mushrooms
1 packet enoki mushrooms
1 broccoli
2 tsps minced garlic
1 tsp minced ginger

芡汁 | Thickening

上湯 1/3 杯
蠔油 3 湯匙
生粉 2 茶匙
老抽 1 茶匙
糖 1 茶匙
鹽 1/4 茶匙

1/3 cup chicken broth
3 tbsps oyster sauce
2 tsps cornstarch
1 tsp dark soy sauce
1 tsp sugar
1/4 tsp salt

做法 | Method

1. 材料洗淨。把本菇和金菇去掉莖部，稍沖水備用。
2. 西蘭花去掉硬皮，汆水，過冷，備用。
3. 熱鑊下油 1 湯匙，爆香蒜茸和薑茸，放入西蘭花兜炒片刻，潷酒，加入 1~2 湯匙上湯，盛起。
4. 將蠔油芡汁倒入鑊中煮滾，放入本菇和金菇煮至軟，盛起。
5. 雜菌與西蘭花同上碟，伴以鍋巴享用。

1. Rinse ingredients. Remove the stems of hon shimeji mushrooms and enoki mushrooms, rinse slightly and set aside.
2. Remove rough skin of broccoli, blanch, rinse and set aside.
3. Heat wok with 1 tbsp of oil, sauté minced garlic and ginger until fragrant. Add broccoli and stir-fry for a while, sprinkle wine, add 1-2 tbsps of broth and set aside.
4. Pour oyster sauce ingredients into wok and bring to a boil. Add hon shimeji and enoki mushrooms, cook until softened and set aside.
5. Dish up assorted mushrooms and broccoli and serve with crispy rice.

海鮮荷葉飯

Seafood Rice in Lotus Leaf

材料 | Ingredients

白飯 400 克
蝦仁 4 隻
冬菇 3 朵
青口 3 隻
魚柳 1 塊
荷葉 1 張

400g cooked white rice
4 shelled shrimps
3 dried black mushrooms
3 mussels
1 fish fillet
1 sheet lotus leaf

2 人
Serves 2

30 分鐘
30 minutes

調味料 | Seasonings

生抽 1 茶匙	1 tsp light soy sauce
熟油 1 茶匙	1 tsp cooked oil
麻油 1/2 茶匙	1/2 tsp sesame oil
胡椒粉 1/4 茶匙	1/4 tsp pepper

做法 | Method

1. 材料洗淨。荷葉放滾水中浸軟，抹乾。
2. 冬菇浸軟，去蒂，切粒。
3. 魚柳、青口和蝦仁分別切粒。
4. 煮滾一鍋水，加入冬菇、魚柳、青口和蝦粒煮熟，盛起，加入調味料拌勻。
5. 將荷葉攤平，放上白飯，加入各材料，裹成四角形包好，放蒸籠內，隔水蒸 12 分鐘即成。

1. Rinse ingredients. Soak lotus leaf in hot water until soft and pat dry.
2. Soak dried black mushrooms, remove stems and dice.
3. Dice fish fillet, mussels and shrimps respectively.
4. Bring a pot of water to a boil, add dried black mushrooms, fish fillet, mussels and shrimps and cook until done, drain and mix with seasonings.
5. Place lotus leaf onto the table flatly, add cooked rice and other ingredients and wrap to form square shape. Place in a steamer and steam for 12 minutes. Serve.

入廚貼士 | Cooking Tips

- 新鮮荷葉有季節性，乾的荷葉比較容易購買，可在雜貨店購買。
- Fresh lotus leaf is a seasonal product. It is easier to buy dried lotus leaf in groceries

有錢佬炒飯

Richman's Fried Rice

◯◯◯ 材料 | Ingredients

帶子 4 粒	本菇 30 克
蝦 6 隻	紅甜椒 1/2 個
雞髀菇 40 克	白飯 3 碗
蘆筍 30 克	蒜茸 1 湯匙

4 scallops

6 shrimps

40g chicken spleen mushrooms

30g asparagus

30g hon shimeji mushrooms

1/2 red bell pepper

3 bowls cooked white rice

1 tbsp minced garlic

海鮮
Seafood

⬤⬤ 醃料 | Marinade

生粉 1/2 茶匙
鹽 1/4 茶匙
胡椒粉 1/4 茶匙

1/2 tsp cornstarch
1/4 tsp salt
1/4 tsp pepper

⬤⬤ 調味料 | Seasonings

蠔油 1 湯匙
XO 醬 1 茶匙
麻油 1/2 茶匙

1 tbsp oyster sauce
1 tsp XO sauce
1/2 tsp sesame oil

⬤⬤ 做法 | Method

1. 材料洗淨。本菇、雞髀菇、蘆筍切粒，汆水。

2. 紅甜椒去籽，切粒。

3. 蝦去殼去腸。帶子瀝乾水分。蝦和帶子加醃料拌勻。

4. 燒熱油鑊，爆香蒜茸，下本菇、雞髀菇、蘆筍和紅甜椒略炒，再加入蝦和帶子炒至全熟，盛起。

5. 再燒熱油鑊，加入白飯炒至鬆，再加入以上材料和調味料拌勻即成。

1. Rinse ingredients. Dice and blanch hon shimeji mushrooms, chicken spleen mushrooms and asparagus.

2. Remove seeds and dice red bell pepper.

3. Remove shells and intestines from shrimps. Drain scallops. Marinate shrimps and scallops.

4. Heat wok with oil. Sauté minced garlic until fragrant, add hon shimeji mushrooms, chicken slpeen mushrooms, asparagus and red bell pepper and stir-fry for a while, then add shrimps and scallops and stir-fry until well done.

5. Heat wok with oil again. Add cooked rice and stir-fry until loosen. Add above ingredients and seasonings and stir-fry well. Serve.

> 入廚貼士 | Cooking Tips
> * 帶子可用雪藏或新鮮的。
> * You may use frozen or fresh scallops.

紅咖喱海鮮炒飯

Fried Rice with Seafood in Red Curry

材料 | Ingredients

青口 4 隻　　　白飯 3 碗
蝦仁 4 隻　　　紅咖喱醬 2 茶匙
魚柳 1 條　　　蒜茸 1 茶匙
芥蘭梗 2 棵

4 mussels
4 shelled shrimps
1 fish fillet
2 Chinese kale stems
3 bowls cooked rice
2 tsps red curry paste
1 tsp minced garlic

4 人
Serves 4

20 分鐘
20 minutes

醃料 | Marinade

鹽少許
胡椒粉少許
生粉少許

some salt
some pepper
some cornstarch

調味料 | Seasonings

生抽 1/2 茶匙
鹽 1/2 茶匙
糖 1/4 茶匙

1/2 tsp light soy sauce
1/2 tsp salt
1/4 tsp sugar

做法 | Method

1. 材料洗淨。蝦仁、青口、魚柳分別切粒。汆水，瀝乾水分。加入醃料拌勻。
2. 芥蘭梗切粒，汆水。
3. 燒熱油鑊，爆香蒜茸和紅咖喱醬，加入白飯炒至鬆散熱透，下調味料拌勻，再加入蝦仁、青口、魚柳略炒，下芥蘭粒炒勻即可上碟。

1. Rinse ingredients. Dice shrimps, mussels and fish fillet respectively. Blanch, drain and marinate.
2. Dice and blanch Chinese kale stems.
3. Heat wok with oil and sauté minced garlic and red curry paste, add cooked rice and stir- fry until loosen. Then add seasonings, shrimps, mussels and fish fillet and stir-fry for a while, add Chinese kale stems and mix well. Serve.

入廚貼士 | Cooking Tips

- 芥蘭可改為青、紅椒。
- You may use green and red bell pepper to substitute Chinese kale.

日式海鮮炒飯

Fried Rice with Seafood in Japanese Style

⟨⟨⟩⟩ 材料 | Ingredients

煙三文魚 100 克
鰻魚 100 克
蝦仁 100 克
蟹子 3 湯匙

雞蛋 1 隻
白飯 3 碗
香草少許

100g smoked salmon
100g eel
100g shelled shrimps
3 tbsps crab roe
1 egg
3 bowl cooked white rice
some herbs

⊙⊙ 調味料 | Seasonings

鰻魚燒汁 3 茶匙	3 tsps eel BBQ sauce
鹽 1/2 茶匙	1/2 tsp salt
糖 1/2 茶匙	1/2 tsp sugar
胡椒粉 1/4 茶匙	1/4 tsp pepper

⊙⊙ 做法 | Method

1. 材料洗淨。蝦、煙三文魚和鰻魚分別切粒。

2. 雞蛋去殼，放碗中打成蛋液。

3. 燒熱油鑊，加入蝦仁炒熟，再下煙三文魚、鰻魚、香草和調味料拌勻，盛起。

4. 再燒熱油鑊，下蛋液炒散，加入白飯炒至飯熱透，將炒好的海鮮回鑊，加入蟹子拌勻即可上碟。

1. Rinse ingredients. Dice shrimps, smoked salmon and eel respectively.
2. Remove egg shell and whisk in a large bowl.
3. Heat wok with oil. Add shrimps and stir-fry until well done. Then add smoked salmon, eel, herbs and seasonings and mix well, dish up.
4. Heat wok with oil again. Add whisked egg and stir-fry until loosen. Add cooked white rice and cook until well done, return cooked seafood and add crab roe, mix well. Serve.

入廚貼士 | Cooking Tips

- 香草可用千里香或雜香草。
- You may use thyme or mixed herbs.

蜆肉拌飯

Rice with Clams

2 人
Serves 2

15 分鐘
15 minutes

材料 | Ingredients

蜆肉 200 克
白飯 2 碗
蒜茸 2 湯匙
急凍青豆 1 湯匙

200g shelled clams
2 bowls cooked white rice
2 tbsps minced garlic
1tbsp frozen green beans

調味料 | Seasonings

忌廉湯 100 毫升
酒 1/2 湯匙
水 120 毫升

100ml cream soup
1/2 tbsp wine
120 ml water

做法 | Method

1. 材料洗淨。蜆肉瀝乾水分。
2. 青豆汆水，瀝乾水分。
3. 白飯放在深碟。
4. 燒熱油鑊，爆香蒜茸，加入蜆肉炒勻，灒酒，加入青豆、忌廉湯和水，煮滾，淋在飯上即可。

1. Rinse ingredients. Drain clams.
2. Blanch green beans and drain.
3. Put cooked white rice onto a deep plate.
4. Heat wok with oil. Sauté minced garlic, add clams and stir well, sprinkle wine. Add green beans, cream soup and water and cook until well done. Spread onto rice and serve.

入廚貼士 | Cooking Tips

- 蜆肉可購買新鮮的，但買回來要先用水浸 2 小時，令蜆肉吐出沙，再汆水，然後取出蜆肉。
- If fresh clams are used, soak for 2 hours to remove sand, then blanch and remove clam shells.

4 人
Serves 4

30 分鐘
30 minutes

鹹魚肉餅飯

Rice with Steamed Minced Pork with Salted Fish

◯◯◯ 材料 │ Ingredients

白米 300 克
半肥瘦豬肉 250 克
鹹魚 20 克
薑 4 片
油少許

300g white rice
250g semi-fat pork
20g salted fish
4 slices ginger
some oil

⟨⟨⟩⟩ 醃料 | Marinade

胡椒粉 1/2 茶匙
鹽 1/3 茶匙

1/2 tsp pepper
1/3 tsp salt

⟨⟨⟩⟩ 調味料 | Seasonings

老抽 1 湯匙
生抽 1 茶匙

1 tbsp dark soy sauce
1 tsp light soy sauce

⟨⟨⟩⟩ 做法 | Method

1. 材料洗淨。白米淘洗淨，瀝乾水分。
2. 薑用小刀刮去皮，切絲。鹹魚瀝乾水分。
3. 半肥瘦豬肉幼切粗剁，放入大碗中，加入醃料，拌至起膠，加入少許油。
4. 白米放電飯煲中，加入適量水煮，當飯將乾水時將半肥瘦豬肉放在飯面，加上鹹魚和薑絲，蓋上蓋焗至全熟即可。
5. 食用時淋上調味料。

1. Rinse ingredients. Rinse white rice and drain.
2. Peel ginger and shred. Drain salted fish.
3. Slice and chop semi-fat pork, put in a large bowl, add marinade and mix until sticky, stir in some oil.
4. Cook rice in rice cooker with appropriate amount of water. Add semi-fat pork when little water left. Add salted fish and shredded ginger, cover until well done.
5. Sprinkle seasonings and serve.

入廚貼士 | Cooking Tips

- 煮熟的飯不要立即開蓋，要焗片刻才會有香味。
- Do not open cover after rice is cooked immediately. Cover for a while to give aroma.

三色有味飯

材料 | Ingredients

臘腸 2 條　　　　黑糯米 100 克
蝦米 30 克　　　　葱 2 條
乾茶樹菇 50 克　　上湯 300 毫升
白米 150 克　　　水適量
糯米 100 克

2 preserved sausages
30g dried shrimps
50g dried tea tree mushrooms
150g white rice
100g brown rice
100g black glutinous rice
2 stalks spring onion
300ml chicken broth
some water

4 人
Serves 4

45 分鐘
45 minutes

調味料 | Seasonings

生抽 2 湯匙	2 tbsps light soy sauce
老抽 1 湯匙	1 tbsp dark soy sauce
油 2 茶匙	2 tsps oil
糖 1/2 茶匙	1/2 tsp sugar

做法 | Method

1. 材料洗淨。糙米、黑糯米預先用水浸 2 小時。

2. 三種米淘洗淨。將三種米拌勻,加入上湯和適量水,放電飯煲內煲熟。

3. 臘腸隔水蒸 10 分鐘至熟,切粒。蝦米、乾茶樹菇浸軟,瀝乾水分。

4. 燒熱油鑊,爆香蝦米、臘腸、乾茶樹菇,加少許鹽拌勻,盛起。

5. 再燒熱油鑊,將調味料拌勻,煮滾備用。

6. 將炒好的材料和調味料加入熟飯拌勻即可。

1. Rinse ingredients. Soak brown rice and black glutinous rice for 2 hours respectively.
2. Rinse three types of rice and mix well. Put into a rice cooker, add chicken broth and some water and cook until done.
3. Steam preserved sausages for 10 minutes until well done and dice. Soak dried shrimps and dried tea tree mushrooms, drain.
4. Heat wok with oil. Sauté dried shrimps, preserved sausages and dried tea tree mushrooms, add some salt and stir well. Set aside.
5. Heat wok with oil again. Stir in seasonings, bring to a boil and set aside.
6. Mix stir-fried ingredients and seasonings with rice. Serve.

入廚貼士 | Cooking Tips

- 菇類可改用新鮮茶樹菇、冬菇或其他菇類。
- You may use fresh tea tree mushrooms, black mushrooms or any other types of mushrooms.

豉汁排骨鳳爪飯

Steamed Rice with Sparerib,
Chicken Feet in Black Bean Sauce

⊙⊙⊙ 材料 | Ingredients

炸鳳爪 4 隻　　豆豉 3 茶匙
腩排 300 克　　蒜茸 1 茶匙
白米 450 克　　水適量
紅辣椒 1 隻

4 pcs deep-fried chicken feet
300g sparerib
450g white rice
1 red chili
3 tsps black beans
1 tsp minced garlic
Some water

調味料 | Seasonings

鹽 1/4 茶匙	1/4 tsp salt
糖 1/4 茶匙	1/4 tsp sugar
胡椒粉 1/4 茶匙	1/4 tsp pepper
麻油 1/4 茶匙	1/4 tsp sesame oil

汁料 | Sauce

老抽適量
熟油適量

Some dark soy sauce
Some cooked oil

做法 | Method

1. 材料洗淨。白米淘洗淨，瀝乾水分。
2. 鳳爪切去指尖。腩排切件。
3. 豆豉剁茸。紅辣椒去籽，切絲。
4. 燒熱油鑊，爆香蒜茸和豆豉，加入腩排、鳳爪和調味料拌勻。
5. 白米放電飯煲中，加入適量水煮，飯將乾水時將材料放下，蓋好蓋焗熟，食用時加入汁料。

1. Rinse ingredients. Wash white rice, and drain.
2. Remove fingertips from chicken feet. Chop spareribs into pieces.
3. Chop black bean into purée. Remove seeds of red chili and shred.
4. Heat wok with oil. Sauté minced garlic and black beans, add spareribs, chicken feet and seasonings and mix well.
5. Cook white rice in rice cooker with appropriate amount of water. Add ingredients when little water left. Cover until well done. Add sauce when serve.

入廚貼士 | Cooking Tips

- 可將鳳爪、腩排和調味料直接放在飯面，這樣可節省時間，但味道會比較遜色。
- Place chicken feet, spareribs and seasonings onto white rice to cook together can save time but it is less tasty.

生炒牛肉飯

Stir-fried Rice with Beef

◯◯◯ 材料 | Ingredients

牛肉 250 克
雞蛋 1 隻
白飯 4 碗
葱 3 條
蒜茸 1 湯匙

250g beef
1 egg
4 bowls cooked white rice
3 stalks spring onion
1 tbsp minced garlic

4 人
Serves 4

20 分鐘
20 minutes

⊗ 醃料 | Marinade

生油 1 湯匙（後下）
生抽 1 茶匙
生粉 1 茶匙
胡椒粉 1/4 茶匙
水 1 茶匙

1 tbsp oil (add at last)
1 tsp light soy sauce
1 tsp cornstarch
1/4 tsp pepper
1 tsp water

⊗ 調味料 | Seasonings

生抽 1 茶匙
老抽 1 茶匙
鹽 1/2 茶匙

1 tsp light soy sauce
1 tsp dark soy sauce
1/2 tsp salt

入廚貼士 | Cooking Tips

- 不可以用鹽醃牛肉，否則會肉質會變靱。
- Do not use salt to marinate beef, otherwise the texture will be tough.

⊗ 做法 | Method

1. 材料洗淨。牛肉剁碎，加醃料拌勻，下生油蓋面。
2. 葱切去根部和尾部，切粒。
3. 雞蛋去殼，打成蛋液。
4. 燒熱油鑊，爆香蒜茸，下牛肉炒至 5 成熟，盛起。
5. 再燒熱油鑊，下白飯和蛋液炒透，將牛肉回鑊，加葱粒和調味料炒勻即可上碟。

1. Rinse ingredients. Mince beef, marinate and add oil on top.
2. Remove roots and end part from spring onion, dice.
3. Remove egg shell and whisk egg.
4. Heat wok with oil. Sauté minced garlic, add beef and stir-fry until medium rare.
5. Heat wok with oil again. Add cooked white rice and whisked egg. Return beef, add diced spring onion and seasonings, stir well and serve.

生炒糯米飯

Stir-fried Glutinous Rice

⦿ 材料 | Ingredients

臘肉 1 條	1 pc preserved pork
臘腸 3 條	3 preserved sausages
蝦米 80 克	80g dried shrimps
糯米 600 克	600g glutinous rice
芫荽 1 棵	1 stalk coriander
葱 1 條	1 stalk spring onion
水約 1 杯	1 cup water

⦿ 調味料 | Seasonings

雞湯 1 杯	1 cup chicken broth
鹽 1 茶匙	1 tsp salt
老抽 1 茶匙	1 tsp dark soy sauce
酒 1 茶匙	1 tsp wine
麻油 1/2 茶匙	1/2 tsp sesame oil

◯◯◯ 做法 | Method

1. 材料洗淨。糯米用水浸 2 至 3 小時,淘洗淨,瀝乾水分。

2. 臘肉、臘腸瀝乾水分。隔水蒸 10 分鐘,切粒。

3. 蝦米浸軟,瀝乾水分。

4. 燒熱油鑊,爆香臘肉、臘腸和蝦米,加麻油拌勻,灒酒,盛起備用。

5. 再燒熱鑊,下油 6 至 7 湯匙,加入糯米輕力炒鬆,炒至有黏性。將水逐少灑下,炒至糯米 8 成熟,再逐少灑下雞湯至糯米全熟,下蒸好的臘肉、臘腸、蝦米、芫荽、葱和老抽拌勻即可。

1. Rinse ingredients. Soak glutinous rice for 2-3 hours, rinse and drain.

2. Drain preserved pork and sausages, drain. Steam for 10 minutes and dice.

3. Soak dried shrimps and drain.

4. Heat wok with oil. Sauté preserved pork, sausages and dried shrimps until fragrant, stir in sesame oil and sprinkle wine. Set aside.

5. Heat wok with 6-7 tbsps of oil, add glutinous rice and stir-fry until loosen and sticky. Sprinkle water by batch, stir-fry glutinous rice until almost done. Stir in chicken broth by batch until glutinous rice is well done. Add steamed preserved pork, preserved sausages, dried shrimps, coriander, spring onion and dark soy sauce. Mix well and serve.

入廚貼士 | Cooking Tips

- 糯米大約需要炒 10 至 15 分鐘。可隨意加減水和雞湯的份量。
- The time for stir-frying glutinous rice is around 10-15 minutes. You may add more or less water and chicken broth.

肉鬆薑米蛋白炒飯

Fried Rice with Pork Floss,
Ginger and Egg White

豬瘦肉 80 克
雞蛋白 3 隻
薑米 2 茶匙
葱 4 條
白飯 4 碗

80g lean pork
3 egg whites
2 tsps minced ginger
4 stalks spring onion
4 bowls cooked white rice

3 人
Serves 3

20 分鐘
20 minutes

(○○) 醃料 | Marinade

生粉 1/2 茶匙
鹽 1/4 茶匙
糖 1/4 茶匙

1/2 tsp cornstarch
1/4 tsp salt
1/4 tsp sugar

(○○) 調味料 | Seasonings

鹽 1/2 茶匙
雞粉 1/2 茶匙
老抽 1/4 茶匙

1/2 tsp salt
1/2 tsp chicken powder
1/4 tsp dark soy sauce

(○○) 做法 | Method

1. 材料洗淨。豬瘦肉瀝乾水分,切細粒再剁碎,加入醃料醃 15 分鐘。
2. 蔥切去根部和尾部,切細粒。
3. 雞蛋白放大碗中打成蛋液。
4. 燒熱鑊,下油約 1 湯匙,加入雞蛋白炒至凝固,盛起。
5. 再燒熱油鑊,下豬肉碎和薑米炒至豬肉熟,加入白飯炒勻至熱,下調味料拌勻,將雞蛋白回鑊,灑下蔥粒即可。

1. Rinse ingredients. Drain, dice and mince lean pork, marinate for 15 minutes.
2. Remove root and end part of spring onion, dice.
3. Whisk egg white in a large bowl.
4. Heat wok with 1 tbsp of oil, add egg white and stir-fry until coagulated, dish up.
5. Heat wok with oil again, add minced pork and ginger, stir-fry until pork is well done. Add cooked white rice and stir-fry well. Then add seasonings and mix well, return egg whites, sprinkle diced spring onion and serve.

入廚貼士 | Cooking Tips

- 雞蛋白不可炒得太久,否則不夠滑。
- Do not stir-fry egg white for too long or else the texture will not be smooth enough.

鴛鴦炒飯
Duo-colour Fried Rice

材料 | Ingredients

蟹肉 40 克	40g crab meat
蝦 6 隻	6 shrimps
雞胸肉 1 件	1 pc chicken breast
洋葱 1 個	1 onion
雞蛋 2 隻	2 eggs
白飯 3 碗	3 bowls cooked white rice

調味料 | Seasonings

生粉 1/2 茶匙
鹽 1/4 茶匙
胡椒粉 1/8 茶匙
1/2 tsp cornstarch
1/4 tsp salt
1/8 tsp pepper

雞絲汁料 | Sauce of shredded chicken

茄汁 4 湯匙	4 tbsps ketchup
糖 1 1/2 茶匙	1 1/2 tsps sugar
生粉 2/3 茶匙	2/3 tsp cornstarch
鹽 1/2 茶匙	1/2 tsp salt
胡椒粉 1/4 茶匙	1/4 tsp pepper
水 50 毫升	50 ml water

(∞) 蟹肉汁料 | Sauce of crab meat

鮮奶 120 毫升	120 ml fresh milk
雞蛋白 1 隻	1 egg white
生粉 2/3 茶匙	2/3 tsp cornstarch
鹽 1/2 茶匙	1/2 tsp salt
糖 1/2 茶匙	1/2 tsp sugar
胡椒粉 1/4 茶匙	1/4 tsp pepper

(∞) 做法 | Method

1. 材料洗淨。蟹肉瀝乾水分,加少許鹽略醃。蝦去殼去腸,洗淨,瀝乾水分。雞胸肉切絲。洋葱去皮、切絲。
2. 燒熱油鑊,下蝦和一半調味料炒熟,盛起。
3. 燒熱油鑊,加雞肉和另一半調味料,泡油後盛起備用。再燒熱油鑊,下洋葱炒香,盛起。
4. 雞蛋去殼,放大碗中打成蛋液。燒熱油鑊,下雞蛋液拌炒至熟,盛起切粒。再加入白飯和蛋液同炒勻,上碟。
5. 燒熱油鑊,下雞絲汁料煮至濃稠,混入洋葱和雞絲拌勻,淋在一半飯面上。
6. 再燒熱油鑊,下蟹肉汁料煮至濃稠,加入蟹肉和蝦肉拌勻,淋在另一邊飯面上便成。

1. Rinse ingredients. Drain crab meat and marinate with some salt. Shell and remove intestines from shrimps, rinse and drain. Shred chicken breast. Peel and shred onion.
2. Heat wok with oil, add 1/2 portion of shrimps and seasonings and stir-fry, dish up.
3. Heat wok with oil, add 1/2 portion of chicken breast and seasonings, parboil and set aside. Heat wok with oil again and sauté onion, dish up.
4. Remove egg shells and whisk eggs. Heat wok with oil. Stir-fry eggs until well done and dice. Add rice and whisked egg and stir-fry, dish up.
5. Heat wok with oil, add sauce of shredded chicken and cook until thickened. Add onion and shredded chicken and stir-fry well. Pour onto half of the rice.
6. Heat wok with oil again, add sauce of crab meat and cook until thickened. Add crab meat and shrimp meat. Pour onto another side of rice.

肉類 Meat

55

福建炒飯

Fried Rice in Fujian Style

⊗ 材料 | Ingredients

雞肉 120 克	120g chicken
燒鴨肉 120 克	120g roasted duck meat
蝦 8 隻	8 shrimps
帶子 4 粒	4 scallops
冬菇 4 朵	4 dried black mushrooms
雞蛋 2 隻	2 eggs
芥蘭梗 2 棵	2 stalks Chinese kale stems
白飯 3 碗	3 bowls cooked white rice

⊗ 海鮮醃料 | Seafood Marinade

生粉 1/2 茶匙
鹽 1/4 茶匙
胡椒粉 1/4 茶匙

1/2 tsp cornstarch
1/4 tsp salt
1/4 tsp pepper

⊗ 雞肉醃料 | Chicken Marinade

生粉 1/2 茶匙
鹽 1/4 茶匙

1/2 tsp cornstarch
1/4 tsp salt

4 人
Serves 4

20 分鐘
20 minutes

⊗ 調味料 | Seasonings

雞湯 1/2 杯
蠔油 1 茶匙
老抽 1/2 茶匙
胡椒粉 1/4 茶匙

1/2 cup chicken broth
1 tsp oyster sauce
1/2 tsp dark soy sauce
1/4 tsp pepper

⊗ 芡汁 | Thickening

生粉 1 茶匙
水 1 湯匙

1 tsp cornstarch
1 tbsp water

入廚貼士 | Cooking Tips

- 福建炒飯的汁料比較深色，所以要用老抽增加其色素。
- The sauce of Fried Rice in Fujian Style is darker. Therefore, dark soy sauce is used to enhance colour.

⊗ 做法 | Method

1. 材料洗淨。蝦去殼去腸，蝦和帶子加海鮮醃料拌勻。
2. 雞肉洗淨，去皮，切粒，加雞肉醃料拌勻。
3. 芥蘭梗、燒鴨肉分別切粒。
4. 冬菇浸軟，去蒂，切粒。雞蛋去殼，放大碗中打成蛋液。
5. 燒熱油鑊，下白飯和蛋液炒勻，上碟備用。
6. 再燒熱油鑊，加入雞肉粒和冬菇炒至雞肉 8 成熟，加入蝦和帶子拌勻，再加入燒鴨粒和芥蘭粒，然後加入調味料煮滾，最後下芡汁拌勻，淋在炒飯面即可。

1. Rinse ingredients. Shell and remove intestines from shrimps. Marinate shrimps and scallops with seafood marinade.
2. Remove skin from chicken, dice and marinate.
3. Dice Chinese kale stems and roasted duck respectively.
4. Soak dried black mushrooms, remove stems and dice. Remove egg shells and whisk eggs in a large bowl.
5. Heat wok with oil, add white rice and whisked egg, stir-fry well. Set aside.
6. Heat wok with oil again, add chicken meat and dried black mushrooms. Stir-fry until chicken meat is almost well done. Add shrimps, scallops, diced roasted duck meat and Chinese kale, then seasonings, cook for a while, stir in thickening and pour onto rice. Serve.

牛仔骨咖喱炒飯

Fried Rice with Curry Beef Short Rib

材料 | Ingredients

牛仔骨 320 克	320g beef short ribs
番茄 2 個	2 tomatoes
洋蔥 1 個	1 onion
雞蛋 2 隻	2 eggs
急凍紅蘿蔔粒 1 湯匙	1 tbsp frozen diced carrot
急凍青豆 1 湯匙	1 tbsp frozen green beans
白飯 3 碗	3 bowls cooked white rice
蒜茸 2 茶匙	2 tsps minced garlic
咖喱醬 2 茶匙	2 tsps curry paste
雞蛋 1 隻	1 egg

醃料 | Marinade

生抽 1 茶匙	1 tsp light soy sauce
糖 1/2 茶匙	1/2 tsp sugar
黑胡椒 1/2 茶匙	1/2 tsp black pepper
生粉 1/2 茶匙	1/2 tsp cornstarch

調味料 | Seasonings

蠔油 3 茶匙	3 tsps oyster sauce
雞粉 1/2 茶匙	1/2 tsp chicken powder
黑胡椒粉 1/2 茶匙	1/2 tsp black pepper

(◯◯) 做法 | Method

1. 材料洗淨。牛仔骨加醃料拌勻。
2. 洋葱去衣、切絲。番茄切粒。
3. 紅蘿蔔粒、青豆汆水。
4. 雞蛋去殼，放大碗中打成蛋液。
5. 燒熱油鑊，下白飯和蛋液炒勻，盛起備用。再燒熱油鑊，下牛仔骨煎至金黃，盛起備用。
6. 再燒熱油鑊，爆蒜茸和咖喱醬，下洋葱、番茄拌炒，再加入調味料、紅蘿蔔粒和青豆拌勻，將牛仔骨回鑊，加入雞蛋飯拌勻即可。

1. Rinse ingredients. Marinate beef short ribs.
2. Peel and shred onion. Rinse tomatoes and dice.
3. Blanch diced carrot and green beans.
4. Remove egg shells and whisk in a large bowl.
5. Heat wok with oil. Add cooked white rice and whisked eggs, set aside. Heat wok with oil again, shallow-fry beef short ribs and set aside.
6. Heat wok with oil, sauté minced garlic and curry paste. Add onion, tomatoes, then seasonings, carrot, green beans, beef short ribs and egg. Stir-fry until well done.

入廚貼士 | Cooking Tips
- 牛仔骨不要煎得太熟。
- Do not overcook beef short ribs.

台式滷肉飯

⚪⚪ 材料 | Ingredients

免治豬肉 300 克
冬菇 4 朵
白飯 3 碗
乾蔥頭 4 粒
蒜頭 2 粒
蝦米 1/2 茶匙

300g minced pork
4 dried black mushrooms
3 bowls cooked white rice
4 shallots
2 cloves garlic
1/2 tsp dried shrimps

3 人
Serves 3

20 分鐘
20 minutes

⟨⟨⟩⟩ 醃料 | Marinade

糖 1/2 茶匙
生抽 1/2 茶匙
生粉 1/2 茶匙
鹽 1/4 茶匙

1/2 tsp sugar
1/2 tsp light soy sauce
1/2 tsp cornstarch
1/4 tsp salt

⟨⟨⟩⟩ 調味料 | Seasonings

老抽 150 毫升　　糖 1/2 茶匙
酒 1 茶匙　　　　水 100 毫升
胡椒粉 1/2 茶匙

150 ml dark soy sauce
1 tsp wine
1/2 tsp pepper
1/2 tsp sugar
100 ml water

⟨⟨⟩⟩ 做法 | Method

1. 材料洗淨。白飯保暖。
2. 免治豬肉加醃料拌勻。
3. 乾葱頭、蒜頭去衣，切茸。
4. 蝦米浸軟，洗淨，切碎。冬菇浸軟，去蒂，切粒。
5. 燒熱油鑊，爆香乾葱茸和蒜茸，加入蝦米和冬菇粒拌炒，加入免治豬肉和調味料煮至汁稍乾，淋在飯上即可。

1. Rinse ingredients. Keep cooked white rice warm.
2. Marinate minced pork.
3. Peel and mince shallots and garlic.
4. Soak dried shrimps and chop. Soak dried black mushrooms, remove stems and dice.
5. Heat wok with oil, sauté minced shallots and garlic. Add dried shrimps and dried black mushrooms and stir-fry. Then add minced pork and seasonings and cook until some sauce left. Pour onto rice and serve.

入廚貼士 | Cooking Tips

- 乾葱頭可以炸香，食用時灑在飯面，增加香味。
- You may deep-fry dried shallots and spread onto rice to enhance flavour.

焗豬扒飯

Baked Pork Chop Rice

⟨⟨⟩ 材料 | Ingredients

豬扒 6 件
紅甜椒 1/2 個
青椒 1/2 個
洋葱 1/2 個
白飯 3 碗
蒜茸 2 茶匙

6 pcs pork chop
1/2 red bell pepper
1/2 green bell pepper
1/2 onion
3 bowls cooked white rice
2 tsps minced garlic

醃料 | Marinade

糖 1 茶匙
生抽 1 茶匙
生粉 1 茶匙
鹽 1/2 茶匙

1 tsp sugar
1 tsp light soy sauce
1 tsp cornstarch
1/2 tsp salt

汁料 | Sauce

蠔油 1 茶匙
喼汁 1/2 茶匙
黑胡椒粉 1/2 茶匙
水 100 毫升

1 tsp oyster sauce
1/2 tsp Worcestershire sauce
1/2 tsp black pepper
100ml water

做法 | Method

入廚貼士 | Cooking Tips

- 豬扒用刀背剁一面已可以，剁過的豬扒就不會韌。
- Hammer one side of pork chop with the back of knife blade is enough and the pork chop will be more tender.

1. 材料洗淨。豬扒洗淨，用刀背剁鬆，加入醃料拌勻。
2. 洋葱去衣，切塊。紅椒、青椒分別去籽，切塊。
3. 燒熱油鑊，下豬扒煎至兩面金黃，盛起備用。
4. 白飯放在玻璃深碟，鋪上豬扒。
5. 再燒熱油鑊，爆香蒜茸和洋葱，加入紅甜椒和青椒拌勻，加入汁料煮片刻，淋在豬扒上，放入預熱 180℃的焗爐內，焗至金黃色即可。

1. Rinse ingredients. Rinse pork chop and hammer with the back of knife blade, then marinate.
2. Peel and cut onion into pieces. Remove seeds from red and green bell peppers and cut into pieces.
3. Heat wok with oil, shallow-fry pork chop until golden brown on both sides. Dish up.
4. Put white rice onto a deep glass plate and place pork chop on top.
5. Heat wok with oil and sauté minced garlic and onion. Add red and green bell peppers, stir in sauce and cook for a while and pour onto pork chop. Bake in preheated oven at 180℃ until golden brown.

韓式石頭鍋飯

Dolsot Bibimbap in Korean Style

⟨⟨⟩⟩ 材料 | Ingredients

薄切牛肉 200 克
青瓜 1/2 條
紅蘿蔔 1/3 條
韓國泡菜 50 克
大豆芽菜 40 克
本菇 30 克
雞蛋黃 1 隻
白飯 2 碗

200g thinly sliced beef
1/2 cucumber
1/3 carrot
50g kimchi
40g soybean sprouts
30g hon shimeji mushrooms
1 egg yolk
2 bowls cooked white rice

2 人
Serves 2

20 分鐘
20 minutes

(○○) 醃料 | Marinade

生抽 1 茶匙
胡椒粉 1/4 茶匙

1 tsp light soy sauce
1/4 tsp pepper

(○○) 調味料 | Seasonings

韓式辣椒醬 1 湯匙
生抽 1 茶匙
麻油 1/2 茶匙

1 tbsp Korean chili paste
1 tsp light soy sauce
1/2 tsp sesame oil

(○○) 做法 | Method

1. 材料洗淨。牛肉瀝乾水分,加醃料拌勻。
2. 本菇、大豆芽菜瀝乾水分。青瓜、紅蘿蔔瀝乾水分,去皮,切絲。
3. 燒熱油鑊,煎香牛肉備用。
4. 再燒熱油鑊,下紅蘿蔔、青瓜、本菇、大豆芽菜、韓國泡菜和調味料拌勻,盛起。
5. 白飯放入石頭鍋,飯面鋪上炒好的牛肉和其他材料。
6. 將石頭鍋直接放在爐上加熱,直至有不斷的啪啪聲傳出,以及有少許飯焦的氣味即可熄火,加上雞蛋黃即可。

1. Rinse ingredients. Drain beef and marinate.
2. Drain hon shimeji mushrooms and soybean sprouts. Drain cucumber and carrot, peel and shred.
3. Heat wok with oil. Shallow-fry beef and set aside.
4. Heat wok with oil again. Add carrot, cucumber, hon shimeji mushrooms, soybean sprouts, kimchi and seasonings and stir well, dish up.
5. Put cooked white rice into stone pot. Spread stir-fried beef and other ingredients on top.
6. Heat stone pot over stove directly. Cook until clattering sound is heard and rice is slightly burnt. Add egg yolk on top and serve.

入廚貼士 | Cooking Tips

- 石頭鍋新買回來要先用水浸 1 小時以上。
- Soak new stone pot in water over 1 hour before use.

迷你糯米雞

Mini Steamed Glutinous Rice with Chicken

⊙ 材料 | Ingredients

雞肉 1 塊　　1pc chicken meat
叉燒 60 克　　60g barbecued pork
冬菇 2 朵　　2 dried black mushrooms
糯米 400 克　400g glutinous rice
荷葉 2 塊　　2 sheets lotus leaf

⊙ 調味料 | Seasonings

蠔油 1 湯匙
水 120 毫升

1 tbsp oyster sauce
120ml water

⊙ 雞肉醃料 | Marinade of chicken

生粉 1/2 茶匙
鹽 1/4 茶匙
胡椒粉 1/4 茶匙

1/2 tsp cornstarch
1/4 tsp salt
1/4 tsp pepper

⟨⟨⟨ 糯米調味料 |
Seasonings of glutinous rice

雞粉 1 茶匙
鹽 1/2 茶匙
麻油 1/2 茶匙
糖 1/4 茶匙
1 tsp chicken powder
1/2 tsp salt
1/2 tsp sesame oil
1/4tsp sugar

⟨⟨⟨ 芡汁 | Thickening

生粉 1 茶匙
水 2 湯匙
1 tsp cornstarch
2tbsps water

⟨⟨⟨ 做法 | Method

1. 材料洗淨。糯米用水浸 3 小時，淘洗淨，瀝乾水分。加入糯米調味料拌勻，放入蒸籠，以大火隔水蒸 20 分鐘，期間翻動糯米飯 2 至 3 次。
2. 荷葉用沸水浸泡後，抹乾水分，剪裁成 8 份。
3. 雞肉切粒，加醃料拌勻。
4. 冬菇浸軟，去蒂，切粒，加少許生抽調味。
5. 燒熱油鑊，將雞肉和冬菇炒熟，加入叉燒和調味料拌勻，下芡汁炒勻，盛起，分成 8 份。
6. 先將一塊荷葉放枱面，先鋪上一層糯米，放入一份材料，再蓋上一層糯米，將荷葉包緊成方形，封口向下，排放在蒸籠上，隔水蒸 8 至 10 分鐘即可。

1. Rinse ingredients. Soak glutinous rice for 3 hours, rinse and drain. Stir in seasoning of glutinous rice and mix well. Put into a steamer and steam for 20 minutes, and flip it over in the steamer 2-3 times.
2. Soak lotus leaf in hot water, pat dry, and cut into 8 portions.
3. Dice and marinate chicken.
4. Soak dried black mushrooms, remove stems, dice and mix with some light soy sauce.
5. Heat wok with oil. Stir-fry chicken and dried black mushrooms, add barbecued pork and seasonings, stir in thickening and divide into 8 portions.
6. Place a piece of lotus leaf onto the table, add some glutinous rice, put a portion of ingredients on top, then add glutinous rice, wrap with lotus leaf to form a square. Place opening downwards, arrange in a steamer and steam for 8-10 minutes.

粢飯

Glutinous Rice Roll

◯◯ 材料 | Ingredients

糯米 300 克
上湯（浸過米的份量）
水 1 1/2 杯

300g glutinous rice
broth (the amount should
be enough to soak rice)
1 1/2 cups water

3 人
Serves 3

60 分鐘
60 minutes

⨀ 餡料 | Fillings

榨菜 1/2 個	1/2 pc preserved vegetables
蝦米 30 克	30g dried shrimps
豬肉鬆 35 克	35g pork floss
油炸鬼 1 條	1 deep-fried twisted dough stick

⨀ 做法 | Method

1. 材料洗淨。糯米淘洗淨，注入上湯和水浸約 3 小時，隔去水分，放鑊中隔水蒸約 30 分鐘至熟，要不時翻動。
2. 榨菜用水略浸，減去鹹味，切碎。
3. 蝦米浸透，切幼粒。
4. 燒熱油鑊，爆香榨菜和蝦米，盛起備用。
5. 放一塊濕布在枱上，鋪上保鮮紙，放上糯米飯鋪平，放入餡料，捲實成飯糰，兩頭紮實即成。

1. Rinse ingredients. Rinse glutinous rice, add in broth and soak for 3 hours. Drain. Steam for 30 minutes until well done. Flip it over continuously.
2. Soak preserved vegetables in water to reduce salty flavour, dice.
3. Soak dried shrimps and dice.
4. Heat wok with oil. Sauté preserved vegetables and dried shrimps. Set aside.
5. Place a piece of wet cloth on table, then line with a sheet of plastic cling. Put glutinous rice on top flatly, then add fillings on top. Roll into rice dumplings in rectangular shape. Tight both ends and serve.

入廚貼士 | Cooking Tips

- 如想改為甜食，餡料可用油炸鬼（油條）和糖。
- If sweet flavour is preferred, you may use deep-fried twisted dough sticks and sugar as fillings.

臘腸蒸雞飯

Steamed Rice with Chicken and Chinese Preserved Sausage

⊘⊘ 材料 | Ingredients

光雞 1/2 隻	1/2 gutted chicken
白米 350 克	350g white rice
臘腸 2 條	2 preserved sausages
葱 2 條	2 stalks spring onion
薑 2 片	2 slices ginger

⊘⊘ 醃料 | Marinade

生抽 1 茶匙	1 tsp light soy sauce
薑汁酒 1 茶匙	1 tsp ginger wine
油 1 茶匙	1 tsp oil
生粉 1/2 茶匙	1/2 tsp cornstarch

⬭ 調味料 | Seasonings

麻油 1 茶匙
老抽適量

1 tsp sesame oil
Some dark soy sauce

⬭ 做法 | Method

1. 材料洗淨。光雞用粗鹽略擦，洗淨，瀝乾水分，切件，加入醃料醃 30 分鐘。
2. 臘腸瀝乾水分。
3. 薑用小刀刮去皮，切絲。葱切去根部和尾部，切粒。
4. 白米淘洗淨，與臘腸同放電飯煲中，加入適量水煮，飯將乾水時下雞件和薑絲，蓋好焗至熟。
5. 食用時將臘腸切片，加入調味料和葱粒。

1. Rinse ingredients. Rub chicken with coarse salt for a while, rinse and drain. Chop into pieces and marinate for 30 minutes.
2. Drain preserved sausages.
3. Peel and shred ginger. Remove root and end part of spring onion and dice.
4. Rinse rice and cook with preserved sausages in rice cooker with appropriate amount of water. Add chicken pieces and shredded ginger when little water left. Cover and cook until well done.
5. Slice preserved sausages, add in seasonings and diced spring onion when serve.

入廚貼士 | Cooking Tips

- 可加入冬菇與雞件同下。
- You may add dried black mushrooms and chicken pieces together.

北菇滑雞片飯

Streamed Rice with Dried Black Mushrooms and Chicken

材料 | Ingredients

雞肉 250 克
冬菇 4 朵
薑 4 片
白米 350 克
250g chicken
4 dried black mushrooms
4 slices ginger
350g white rice

3 人
Serves 3

30 分鐘
30 minutes

◯◯◯ 醃料 | Marinade

生抽 1 茶匙
薑汁酒 1 茶匙
油 1 茶匙
胡椒粉 1/4 茶匙

1 tsp light soy sauce
1 tsp ginger wine
1 tsp oil
1/4 tsp pepper

◯◯◯ 調味料 | Seasonings

熟油 1 茶匙
老抽適量

1 tsp cooked oil
Some dark soy sauce

入廚貼士 | Cooking Tips

- 冬菇浸軟後不用去乾水分，即可切絲，將帶有水分的冬菇一併和雞片放在飯面。
- No need to drain soaked dried black mushrooms completely before shredding. Just put soaked mushrooms and sliced chicken on top of rice.

◯◯◯ 做法 | Method

1. 材料洗淨。白米淘洗淨，瀝乾水分。
2. 雞肉切片，加入醃料拌勻。
3. 薑去皮，切絲。冬菇浸軟，去蒂，切絲。
4. 白米放電飯煲中，加入適量水煮，飯將乾水時下雞片、冬菇和薑絲，蓋好焗熟。
5. 食用時加入調味料。

1. Rinse ingredients. Rinse white rice and drain.
2. Slice chicken and marinate.
3. Peel ginger and shred. Soak dried black mushrooms, remove stems and shred.
4. Cook rice in rice cooker with appropriate amount of water. Add sliced chicken, dried black mushrooms and shredded ginger when little water left. Cover and cook until well done.
5. Add seasonings when serve.

海南雞飯

Hainanese Chicken Rice

◯◯◯ 材料 | Ingredients

光雞 1 隻（約 1500 克）
白米 600 克
薑 1 塊
葱 2 條
蒜頭 10 粒

1 gutted chicken (about 1500g)
600g white rice
1 pc ginger
2 stalks spring onion
10 cloves garlic

調味料 | Seasonings

鹽 4 茶匙 +1 茶匙
胡椒粉適量
酒適量
薑汁適量

4 tsps + 1 tsp salt
Some pepper
Some wine
Some ginger juice

做法 | Method

1. 材料洗淨。白米淘洗淨，
   瀝乾水分。

2. 薑去皮，切片。葱切去根
   部和尾部，切度。蒜頭去衣，剁成茸。

3. 光雞用粗鹽略擦，洗淨，瀝乾水分。切去雞油，切粒，留起待用。

4. 燒滾一大鍋水，將光雞、薑和葱放滾水內浸 10 分鐘，直至光雞煮熟。
   撈起，趁熱擦上鹽和胡椒粉，放涼後斬件上碟。浸雞水留起待用。

5. 燒熱油鑊，加入雞油粒煮融，下蒜茸爆香，灒酒及薑汁，倒入已洗淨
   的白米炒勻，下鹽 4 茶匙拌勻。轉放電飯煲內，注入適量浸雞水煮
   至飯熟即可。

入廚貼士 | Cooking Tips

- 煮飯時可加入椰汁。
- You may add coconut milk when cooking rice.
- 雞飯可伴酸菜。做法是將青瓜 1 條切條，用鹽醃片刻，去清水分，加入白醋 3 湯匙和糖 1 湯匙，放雪櫃冷藏即可。
- Sour vegetables accompany chicken rice is common when serve. Shred one piece of cucumber and marinate with salt for a while. Rinse, add in 3 tbsps of white vinegar and 1 tbsp of sugar, then put in fridge.

1. Rinse ingredients. Rinse white rice and drain.

2. Peel ginger and slice. Remove root and end part of spring onion and chop. Peel and mince garlic.

3. Rub chicken with coarse salt for a while, rinse and drain. Reserve chicken fat, dice and set aside.

4. Bring a pot of water to a boil. Boil chicken, ginger and spring onion for 10 minutes until chicken is well done. Take chicken out, rub with salt and pepper, let cool and chop into pieces and arrange onto a serving plate. Keep chicken soup for use.

5. Heat wok with oil. Add chicken fat and cook until dissolved, then add garlic and stir-fry until fragrant. Sprinkle wine and ginger juice, stir in white rice and stir-fry, add 4 tsps of salt. Put in rice cooker with appropriate amount of chicken soup and cook until well done.

西炒飯

Fried Rice in Western Style

◯◯◯ 材料 | Ingredients

腸仔 3 條
青豆 2 湯匙
青椒 1/2 個
洋葱 1/2 個
雞蛋 1 隻
白飯 3 碗

3 pcs sausages
2 tbsps green beans
1/2 green bell pepper
1/2 onion
1 egg
3 bowls cooked white rice

3 人
Serves 3

15 分鐘
15 minutes

◯◯◯ 調味料 | Seasonings

茄汁 4 湯匙	4 tbsps ketchup
鹽 1/2 茶匙	1/2 tsp salt
生抽 1/2 茶匙	1/2 tsp light soy sauce

◯◯◯ 做法 | Method

1. 材料洗淨。腸仔切粒。青豆汆水。
2. 青椒去籽，切粒。洋葱去衣，切粒。
3. 雞蛋去殼，打成蛋液。
4. 燒熱油鑊，下白飯拌炒，加入蛋液炒至鬆，盛起。
5. 再燒熱油鑊，炒香洋葱，加入腸仔粒、青椒粒和青豆拌炒，加入已炒的白飯，最後加調味料拌勻即可上碟。

1. Rinse ingredients. Dice sausages. Blanch green beans.
2. Remove seeds and dice green bell pepper. Peel and dice onion.
3. Remove egg shell and whisk egg.
4. Heat wok with oil, stir-fry cooked white rice, stir in whisked egg until stir-fry until well done.
5. Heat wok with oil again. Sauté onion, add sausages, green bell pepper and green beans, then fried rice and seasonings and mix well. Serve.

入廚貼士 | Cooking Tips

- 下了茄汁，如覺得味道太酸，可加少許糖。
- You may add some sugar if the taste of rice is sour after adding ketchup.

揚州炒飯 Yangzhou Fried Rice

◯◯◯ 材料｜Ingredients

蝦 8 隻
叉燒 120 克
白飯 3 碗
葱 3 條
雞蛋 2 隻

8 shrimps
120g barbecued pork
3 bowls cooked white rice
3 stalks spring onion
2 eggs

醃料 | Marinade

生粉 1/2 茶匙
鹽 1/4 茶匙
胡椒粉 1/4 茶匙

1/2 tsp cornstarch
1/4 tsp salt
1/4 tsp pepper

調味料 | Seasonings

鹽 1 茶匙
生抽 1 茶匙

1 tsp salt
1 tsp light soy sauce

做法 | Method

1. 材料洗淨。蝦去殼去腸，加醃料拌勻。
2. 叉燒切粒，葱切去根部和尾部，切粒。
3. 雞蛋去殼，打成蛋液，加少許鹽拌勻。
4. 燒熱油鑊，下蛋液炒勻，盛起，切絲。
5. 再燒熱油鑊，下蝦仁略炒，加入叉燒粒拌炒，再加入白飯和調味料，最後加入雞蛋絲和葱粒拌勻。

1. Rinse ingredients. Shell and remove intestines from shrimps and marinate.
2. Dice barbecued pork. Remove root and end part of spring onion and dice.
3. Remove egg shells, whisk eggs and stir in some salt.
4. Heat wok with oil, stir in eggs, stir-fry until done and shred.
5. Heat wok with oil again. Add shrimps and barbecued pork and stir-fry. Add white rice and seasonings, and then shredded egg and diced spring onion at last. Serve.

入廚貼士 | Cooking Tips

- 買叉燒時可請店員將叉燒切粒，這樣可省回一些時間。
- You may request the staff to dice barbecued pork to save time.

鹹魚雞粒飯

Fried Rice with Diced Chicken and Salted Fish

材料 | Ingredients

雞肉 250 克　　葱粒 1 湯匙
鹹魚 20 克　　薑 2 片
雞蛋 2 隻　　薑米 1 茶匙
白飯 4 碗　　蒜茸 1 茶匙

250g chicken
20g salted fish
2 eggs
4 bowls cooked white rice
1 tbsp diced spring onion
2 slices ginger
1 tsp minced ginger
1 tsp minced garlic

4 人
Serves 4

20 分鐘
20 minutes

◯◯◯ 醃料 ｜ Marinade	◯◯◯ 調味料 ｜ Seasonings
鹽 1/2 茶匙	生抽 1 茶匙
胡椒粉 1/2 茶匙	老抽 1 茶匙
1/2 tsp salt	鹽 1/2 茶匙
1/2 tsp pepper	1 tsp light soy sauce
	1 tsp dark soy sauce
	1/2 tsp salt

◯◯◯ 做法 ｜ Method

1. 材料洗淨。雞肉切粒，加醃料拌勻。薑切絲。
2. 燒熱油鑊，將雞肉泡油，瀝乾油分備用。
3. 鹹魚瀝乾水分。鹹魚放碟上，鋪上薑絲，隔水蒸 10 分鐘，待涼拆肉。
4. 雞蛋去殼，打成蛋液，加少許鹽拌勻。燒熱油鑊，下蛋液炒勻，盛起，切絲。
5. 燒熱油鑊，爆香薑米和蒜茸，下白飯炒至鬆，將雞肉、鹹魚和雞蛋絲回鑊，加入調味料拌勻，撒上葱粒即可上碟。

1. Rinse ingredients. Dice chicken and marinate. Shred ginger.
2. Heat wok with oil. Parboil chicken and drain.
3. Drain salted fish. Put onto a plate, sprinkle shredded ginger on top, steam for 10 minutes. Let cool and remove bones.
4. Remove egg shells and beat with some salt. Heat wok with oil, add whisked egg and stir well. Dish up and shred.
5. Heat wok with oil. Sauté minced ginger and garlic, add cooked white rice and stir-fry until loosen. Add chicken and salted fish, then seasonings and diced spring onion and stir well. Serve.

入廚貴士 ｜ Cooking Tips
- 鹹魚用梅香會比較甘香。
- "Mui Heung" salted fish taste better.

豉油雞糙米飯

Brown Rice with Soy Sauce Chicken

◯◯◯ 材料 | Ingredients

雞髀 2 隻
糙米 150 克
水 360 毫升

2 chicken thighs
150g brown rice
360 ml water

 汁料 | Sauce

薑 2 片	2 slices ginger
蔥 3 條	3 stalks spring onion
酒 120 毫升	120 ml wine
老抽 1 杯	1 cup dark soy sauce
糖 6 湯匙	6 tbsps sugar
雞精 1/2 粒	1/2 pc essence of chicken

做法 | Method

1. 材料洗淨。雞髀瀝乾水分。蔥切去根部和尾部,切段。

2. 糙米用水浸 30 分鐘以上,淘洗淨,瀝乾水分。

3. 糙米放電飯煲中,加適量水煮至熟。

4. 將汁料放鑊中煮滾,加入雞髀,蓋上鑊蓋,燜焗 20 分鐘至熟,盛起。

5. 將糙米飯盛起,放上雞髀,淋上適量汁料即可。

1. Rinse ingredients. Drain chicken thighs. Remove root and end part of spring onion, cut into sections.
2. Soak brown rice for over 30 minutes, rinse and drain.
3. Cook brown rice in rice cooker with appropriate amount of water until done.
4. Cook sauce in wok, add chicken thighs. Cover and stew for 20 minutes until well done.
5. Put brown rice onto a serving plate and top with chicken thighs. Pour some sauce on top. Serve.

入廚貼士 | Cooking Tips
- 汁料可以用來浸乳鴿或全隻雞。
- The sauce can be used for soaking pigeon or whole chicken.

葡國雞焗飯

Baked Chicken Rice
in Portuguese Style

材料 | Ingredients

光雞 1/2 隻	1/2 gutted chicken
（約 900 克）	(about 900g)
雞蛋 1 隻	1 egg
馬鈴薯 1 個	1 potato
洋葱 1 個	1 onion
乾葱 1 粒	1shallot
白飯 4 碗	4 bowls cooked white rice
咖喱粉 1 茶匙	1 tsp curry powder
咖喱醬 1 茶匙	1 tsp curry paste

醃料 | Marinade

生粉 1 茶匙	1 tsp cornstarch
鹽 1/2 茶匙	1/2 tsp salt
生抽 1/2 茶匙	1/2 tsp light soy sauce
糖 1/4 茶匙	1/4 tsp sugar

3 人
Serves 3

50 分鐘
50 minutes

⓪⓪ 調味料 | Seasonings

生抽 1/2 茶匙
鹽 1/4 茶匙

1/2 tsp light soy sauce
1/4 tsp salt

⓪⓪ 汁料 | Sauce

椰汁 100 毫升
淡奶 50 毫升

100 ml coconut milk
50 ml evaporated milk

⓪⓪ 做法 | Method

1. 材料洗淨。光雞用粗鹽略擦，洗淨，瀝乾水分，切件，加入醃料醃 30 分鐘。雞蛋去殼，放大碗中打成蛋液。
2. 馬鈴薯去皮，切滾刀塊。洋葱去衣，切粗塊。
3. 燒熱油鑊，加入洋葱略炒，下少許鹽調味。
4. 再燒熱油鑊，加入白飯拌炒，加入蛋液炒勻，放入深碟備用。
5. 再下油爆香乾葱、咖喱粉和咖喱醬，加入雞件和薯仔拌勻，下調味料和適量水，蓋好鑊蓋，煮 20 分鐘。再加入洋葱煮約 10 分鐘至各材料已軟脸，加入汁料淋在飯面。
6. 放上已預熱的焗爐內，以 190℃焗約 10 分鐘至金黃色即成。

入廚貼士 | Cooking Tips

- 不喜歡辣可不爆香咖喱，直接加入便可。
- If you dislike spicy flavour, do not sauté curry and add directly.

1. Rinse ingredients. Rub chicken with coarse salt, rinse, drain and chop into pieces. Marinate for 30 minutes. Remove egg shell and whisk egg in a large bowl.
2. Peel potato and cut into wedges. Peel onion and cut into large pieces.
3. Heat wok with oil, sauté onion and stir in some salt.
4. Heat wok with oil again, add white rice and stir-fry, stir in whisked egg and mix well. Set aside.
5. Heat wok with oil. Sauté shallot, curry powder and curry paste. Add chicken and potato, stir in seasonings and some water. Cover and cook for 20 minutes. Add onion and cook for 10 minutes until all ingredients are tender, stir in sauce and pour onto rice.
6. Bake in preheated oven at 190℃ for 10 minutes until golden brown.

火鴨泡飯

Rice with Roasted Duck in Soup

◯◯ 材料 | Ingredients

燒鴨 1/4 隻
菜心 4 條
白飯 2 碗
葱 2 條
1/4 roasted duck
4 stalks choi sum
2 bowls cooked white rice
2 stalks spring onion

調味料 | Seasonings

魚露 1 茶匙	1 tsp fish sauce
雞粉 1/2 茶匙	1/2 tsp chicken powder
老抽 1/4 茶匙	1/4 tsp dark soy sauce

做法 | Method

1. 材料洗淨。燒鴨斬件。
2. 葱切去根部和尾部，切粒。
3. 菜心摘去老葉。
4. 在鑊中煮滾 4 碗水，加入燒鴨和調味料，煮片刻，放下菜心再煮滾，熄火。
5. 白飯保暖，放大碗內，倒入鴨湯，灑上葱粒即成。

1. Rinse ingredients. Chop roasted duck into pieces.
2. Remove root and end part of spring onion and dice.
3. Remove old leaves from choi sum.
4. Boil 4 bowls of water in wok. Add roasted duck and seasonings and cook for a while. Add choi sum and cook until well done.
5. Keep white rice warm and put into a large bowl. Add roasted duck soup and sprinkle diced spring onion and serve.

入廚貼士 | Cooking Tips

- 燒鴨可改用燒鵝或雞，菜可改用任何菜。
- You may use roasted goose or chicken to substitute roasted duck, and use any other kinds of vegetables to substitute choi sum.

粒粒飯香

編著
梁燕

譯者
Sa

攝影
Fanny

編輯
Pheona Tse

美術設計
Venus

排版
劉葉青

出版者
萬里機構出版有限公司
香港鰂魚涌英皇道1065號東達中心1305室
電話：2564 7511
傳真：2565 5539
電郵：info@wanlibk.com
網址：http://www.wanlibk.com
　　　http://www.facebook.com/wanlibk

發行者
香港聯合書刊物流有限公司
香港新界大埔汀麗路36號
中華商務印刷大廈3字樓
電話：2150 2100
傳真：2407 3062
電郵：info@suplogistics.com.hk

承印者
中華商務彩色印刷有限公司
香港新界大埔汀麗路36號

出版日期
二零一九年三月第一次印刷

萬里機構

萬里 Facebook